The Secrets of Words

The Secrets of Words

Noam Chomsky
and Andrea Moro

The MIT Press
Cambridge, Massachusetts
London, England

This book was set in Arnhem Pro by New Best-set Typesetters Ltd. Printed and bound in the United States of America.

Library of Congress Cataloging-in-Publication Data

Names: Chomsky, Noam, interviewee. | Moro, Andrea, interviewer.
Title: The secrets of words / Noam Chomsky and Andrea Moro.
Description: Cambridge, Massachusetts : The MIT Press, [2022] | Includes bibliographical references.
Identifiers: LCCN 2021045982 | ISBN 9780262046718 (hardcover)
Subjects: LCSH: Chomsky, Noam—Philosophy. | Linguistics—Philosophy. | Chomsky, Noam—Interviews. | Linguists—United States—Interviews. | LCGFT: Interviews.
Classification: LCC P85.C47 A5 2022 | DDC 410—dc23/eng/20211203
LC record available at https://lccn.loc.gov/2021045982

10 9 8 7 6 5 4 3 2 1

Contents

The Secrets of Words

AM Hello, Noam. It is very nice to finally see you. We were supposed to have gotten together many months ago: the pandemic prevented it, but at least we're able to meet now, even if remotely. If it's all right with you, I would like to start our conversation with something that really struck me when I was your student at MIT in 1988. In class, you quoted Yehoshua Bar-Hillel on the relationship between technology and language as it was experienced in the early 1950s; it was something he had written some twenty years after the fact. At the time, what he had said was shocking to me, and I think it still pertains in an interesting manner to the contemporary world. If you don't mind, I will just read it out loud before asking for your opinion on it today.

Bar-Hillel, more specifically, was talking about the atmosphere and the ideas circulating at the Research Laboratory of Electronics at MIT in Cambridge, Massachusetts. This is the key passage: "There was an ubiquitous and overwhelming feeling around the Laboratory that with the new insights of cybernetics and the newly developed techniques of information theory the final breakthrough towards a full understanding of the complexities of communication 'in the animal and the machine' had been achieved. Linguists and psychologists, philosophers and sociologists alike hailed the entrance of the electrical engineer and the probability mathematician into the communication field."[1] I would like to ask you if, after so many years, you could again offer your opinion on this.

NC Well, as you remember, Bar-Hillel was saying that the euphoria was misplaced—that what had been confidently anticipated hadn't happened. He was referring to the situation in the early 1950s. That's when Morris Halle, Eric Lenneberg, and I were just getting together and meeting as graduate students at Harvard. And that was in fact the atmosphere at Harvard, and in the Cambridge environment. There was a lot of excitement. It was against a sociopolitical background. It's important to recall that before the Second World War, the United States was kind of an intellectual backwater. If you wanted to study physics, you would go to Germany; if you wanted to study philosophy, you'd go to

England or Vienna; if you wanted to be a writer, you'd go to France. The United States was kind of like some small town, off in the periphery. I mean, there were things happening, great scientists. But it was at the margins.

The war changed all that, totally. Europe was, of course, devastated. The United States gained enormously through the war. Industrial production quadrupled. A lot of scientific discoveries and technological advances were made during the war. After the war, the United States basically owned the world. This seeped into general consciousness: the Europeans were finished, we're taking over. And it was against that background that the new developments Bar-Hillel was describing came to the fore and became part of a feeling that "we're really going forward, we're leaving the old world behind; now we're starting out on a new path, far beyond what anyone else has conceived." This showed up in all kinds of domains, including this one. Now I happened to be located in Cambridge—Harvard and MIT. Claude Shannon's information theory was developed from wartime research. Along with Norbert Wiener's cybernetics, it looked as if a new era was coming. We could now turn to the study of what used to be called in old-fashioned days the "study of mind." But now we will deal with it by the methods of science. And there was enormous excitement.

I should say it is pretty similar today, with the enthusiasm and excitement about artificial intelligence, deep learning: somehow that's going to solve all of the problems. And it's even more problematic. It's not achieving scientific results, in our domains at least. But the atmosphere is very similar.

And I should mention that at the time, around 1950, there was another source, an independent source, of enthusiasm and euphoria: that was American structural linguistics—which in fact wasn't represented in Cambridge. There were no American linguists there. I guess probably I was the only one, thanks to my undergraduate background. Roman Jakobson was there from the European structuralist tradition, but no American structuralists. But elsewhere in the country there was a fairly tightly knit group of structural linguists who had accomplished a good deal, and they felt that, if you read over the material from back then, they had really finally managed to establish linguistics as a science—a "taxonomic science" it was called—for the first time. They had procedures and analyses that were well defined: you could apply them to any language, any materials, and you get the phonemic analysis, the morphology, and a couple words about syntax, but the field was basically done. And in fact, I recall, I was a student at the time, that the feeling among the students was: this is a lot of fun, but what are we going to do when we apply the procedures to every language? Would it be over? And that was the general assumption.

So, there were two separate kinds of consensus, each enthusiastic, and each with the sense that there had been great accomplishments. In one case, the field was pretty much finished; in the other, we were going on to create a new world. It fell apart in the 1950s. Neither worked. With the Cambridge consensus, it was possible, when the proposals were clear enough, to investigate them. For example, the prevailing idea was that language was a Markovian system of production. That was precise: you could prove that it's false. Others were vaguer. You just had more indirect arguments that they were on the wrong track. The structuralist consensus just didn't work. There was a sharp clash, which became more and more prominent, between procedures of analysis, which is what the field was about, and explanatory theories. It turned out that if you tried to develop an explanatory theory—which is what a generative grammar is—if you tried to construct a generative grammar, an explanatory theory, the elements in it could not be reached by the procedures. So, there was a gulf between trying to develop an explanation of phenomena of language and applying procedures of analysis. The projects were inconsistent. What finally prevailed over time was the effort to find explanatory theories.

It is striking to see that we are in a very similar situation today. The excitement coming out from Silicon Valley, basically—lots of hype and propaganda about how amazing the achievements are—has a certain similarity to the technological euphoria that Bar-Hillel was describing. And that, again, by the time he described it, he thought had been undermined, and was a mistake.

AM Your reaction to that passage that referred to the atmosphere of the '50s, which is still important, warns us against the dominating euphoria today. But there is one specific detail that I would like to ask you to reflect on, regarding something else you presented to us in class: it concerns the seminal work that came from a totally different field, namely brain studies and, more specifically, aphasiology. You did so by introducing us to the work of Eric Lenneberg, with his *Biological Foundations of Language*.[2] Do you think that there has been any substantive change in brain studies since the '50s?

NC Oh, it's totally different. Actually, there was interesting work in the field of brain sciences in the early '50s. But interestingly, it was essentially unknown in the areas of psychology, cognitive science, and so on. I'll give you a very striking example of that. Karl Lashley, one of the great neuroscientists, in the late 1940s gave a very important lecture, the Hixon Symposium lecture, it came out in print in 1951. He demonstrated, not so much by brain studies, but just by looking at the nature of the behavior of organisms—horses galloping, people playing music, and so on—that the entire behaviorist framework was hopeless. He showed that it couldn't possibly work for even simple things, like accounting for how a horse can gallop. He gave very strong arguments. You have to recall that at that time, radical behaviorism was central to the euphoric approaches. B. F. Skinner's William James lectures were circulated in 1948, and later came out as his book *Verbal Behavior*. W. V. O. Quine, the most influential philosopher at the time, picked it up, and it was the core of his work. It was at the center: what was central to the sense that we can understand everything was radical behaviorism. Lashley in 1951 knocked the props out from under it.[3] Nobody knew it. I discovered it as a student because it was recommended to me by an art historian, Meyer Schapiro, who was kind of a polymath; he mentioned to me once that I should look at this article. He thought it was interesting. About 1955, I guess. I looked at it: I could see right away it shattered everything. None of the psychologists at Harvard knew it. It's not mentioned in the literature—but in the *neurological* literature, it was mentioned frequently. There it was noticed. Never made it into psychology or cognitive science. I think the first mention of it in these areas was probably in my review of Skinner in 1959.[4] Later it came to be known widely.

Lashley was a major figure in the brain sciences, but his important work didn't enter into psychology or philosophy of language, or the emerging cognitive sciences. Of course, there was other work, like Wilder Penfield's studies, invasive studies of the brain; studies that can't be carried out now for ethical reasons. But the bars were much lower in those days: there were few constraints [*laughs*]. And then Eric

Lenneberg's book came along in 1967. It was really revolutionary. He founded the modern biology of language. And he had very interesting studies on many different topics, including a very interesting chapter on the evolution of language, which to this day remains a classic basis for serious work in the field. He also studied cases of language disability. So, people with virtually no cortex, a virtually undetectable cortex, who had excellent language knowledge. How could that be?

He had earlier made some interesting discoveries, which were considered so impossible he didn't even publish them. We were close personal friends, students together in the early '50s. He was already interested in how language developed with disabilities. So, one of the things he was interested in was language of the deaf. At that time there was a very strict oralist tradition. The deaf were not allowed to learn sign. They had to learn lip-reading. So, parents were instructed not to gesture if you had a deaf child. The schools taught solely lip-reading. Eric went to observe in the most advanced school for the deaf in the Boston area. And he noticed something very interesting. The teacher was teaching the kids with lip-reading, but as soon as the teacher turned to the blackboard, the kids started going like this [*waves hands about*]. And he realized that they must have invented their own sign language. But that idea was considered so outlandish that he never even published it. In much later years, it was discovered that children do indeed invent their own sign language, without input, but this was sixty years too early. He couldn't even publish it back then.

The book was really a breakthrough. However, the techniques weren't yet available to carry out really effective work in the neurosciences that would bear on serious questions as to how language works. The first major breakthrough was in a paper in 2003, the experiments you designed that were carried out in Milan. These were the first ones, and up to this day almost the only ones, that found a really significant basis in the function of the brain for a fundamental property of language: the property called *structure dependence*. A very curious feature of language is that young children, two-year-old children, when they're applying linguistic rules to create and interpret sentences, ignore 100 percent of what they hear, and pay attention only to something they never hear. What they ignore is the linear order of words. That's what you hear. Like, if you listen to us now you hear words coming like beads on a string, one after another. Children totally ignore this in applying the rules of their internal language. What they pay attention to is the structures they are creating in their minds, which of course you never hear. You don't hear a structure. It's just something your mind automatically creates on the basis of a linear ordered string of words. By now there's substantial linguistic evidence about this curious and fundamental property of language, and your experiments, which I'll let you talk about now, show that you can actually find what's going on in the brain that correlates with this. Why don't you take over [*laughs*]?

AM Thank you for giving me this opportunity to describe these experiments and their connection to your hypotheses. Indeed, as you say, the experiment was designed in Milan. In fact, the research paradigm was exploited twice in two separate experiments with two different teams conducted in Germany and Switzerland. The core and common idea was to rely on what you just said, namely that children only pay attention to hierarchical structures rather than linear order. Assuming this, I invented two types of languages: one based on hierarchy (call it a "possible language") and a second one based on linear order (call it an "impossible language"). There are an infinite number of rules that one can invent based on linear order. We basically used three types of rules: rules based on the rigid position of a specific word in the linear sequence of words in a sentence (for example, negation as occurring always as, say, the third word); rules based on a rearrangement of words (for example, a rule that forms an interrogative sentence by composing the mirror image of the sequence of words of the corresponding affirmative sentence); and rules based on agreement between two words at the extremes of a string (for example, the first article agrees with the last noun of a given sentence). It's worth noting that the rule based on rearranging the same words is nothing but an extension of a capacity that is typically used in many languages: take an affirmative sentence like *America is beautiful*; its corresponding interrogative sentence is *Is America beautiful?* In our impossible language, the corresponding rule would give *Beautiful is America* with the same meaning as the former interrogative sentence but as computed on a flat structure rather than on the hierarchical one we find in actual English.

The two experiments were both essentially based on this simple idea of measuring the brain's reaction when judging the grammaticality of sentences based on possible versus impossible rules, but there was also another factor that distinguished them. In one case, we compared real, fully meaningful languages by teaching a micro-version of Italian containing both possible and impossible rules to a group of monolingual German speakers.[5] In the other, we instead let people discover similar rules from sentences that were constructed from a limited vocabulary of invented, though phonologically plausible, words, call them "pseudo-words," preserving only the so-called functional words, such as articles, negation, and auxiliaries. The sentences of this second experiment sounded like jabberwocky and it was impossible to compute a full meaning; for example, with something like *The gulks janidged the brals.*[6] Nevertheless, the speakers learned those rules and the results were substantially comparable to the ones of the first experiment. The reason we also exploited pseudo-words along with normal words was to exclude the possibility that it was not syntax but semantics that enabled the subjects to learn and manage the rules in the impossible languages. Moreover, to refine the first experiment, we taught micro-Japanese along with micro-Italian in order to exclude also the possibility that the similarity between the latter language and the native language of the subjects (both German and Italian belong to the Indo-European languages, whereas Japanese doesn't) could offer any advantage to the speakers. And in fact, there was none. Even the number of errors the subjects made during grammatical tests was similar, independently of whether the language was Italian or Japanese. When we measured the reaction of their brains, the results were robust and sharp.

The conclusion was that even without instructions, the brain is able to recognize possible versus impossible rules—that is, rules based on hierarchy versus rules based on linear order. More specifically, we came to this conclusion by neurobiological measures: the network the brain activated when it employed impossible rules was not the same one it used for possible rules. In fact, the brain circuits that are normally involved in linguistic tasks, which essentially involve a subpart of Broca's area, were progressively *inhibited* when the subjects' accuracy increased as they computed impossible rules; on the other hand, the activity of these same circuits progressively increased when the subjects' accuracy *increased* as they computed possible rules. In fact, impossible rules, that is, linear or "flat" rules, are treated by the brain as a puzzle, involving problem-solving strategies—that is, as something radically different from grammatical structures. In conclusion, "flat tongues" are not human tongues: they can only be spoken in "flat lands."

Of course, these experiments and many subsequent ones in the field, including, among others, those conducted by Angela Friederici, David Poeppel, Alec Marantz, Daniel Osherson, and Stanislas Dehaene, were unconceivable if your proposal that language structure is bounded by neurobiological restrictions was not accepted and language rules rather kept being considered "arbitrary, cultural conventions." These latter words are taken from

Lenneberg's introduction, and I find it worth citing his caveat in full: "A biological investigation into language must seem paradoxical as it is so widely assumed that languages consist of *arbitrary, cultural conventions* [emphasis added]. Wittgenstein and his followers speak of the word game, thus likening languages to the arbitrary set of rules encountered in parlor games and sports. It is acceptable usage to speak of the psychology of bridge or poker, but a treatise on the biological foundation of contract bridge would not seem to be an interesting topic. The rules of natural languages do bear some superficial resemblance to the rules of a game, but I hope to make it obvious in the following chapters that there are major and fundamental differences between rules of languages and rules of games. The former are biologically determined; the latter are arbitrary."[7]

The empirical "proof" that this hypothesis was true was only possible once the scientific community accepted generative grammar and its consequences on language acquisition as the background framework.

More specifically, the experiments on impossible languages I described before could be conceived by strictly relying on your first papers of the '50s, where it was clear, by using mathematical theorems—as opposed to approximate, subjective, and anecdotal intuitions—that statistics could not be enough to fully capture the basic ubiquitous regularities of syntactic structures, such as, for example, nested dependencies. In mathematical terms, you proved that the Markovian chains adopted by Shannon were not sufficient. These experiments offer neurobiological evidence that these regularities are based on a capacity that precedes experience.

NC To me, it seemed that the major significance of your experiment was to find neural correlates to the distinction between possible and impossible languages, focusing on a crucial property: the role of linear order and structures created by the mind. There is compelling evidence that linear order is ignored by the internal language that computes the structures that enter into thought and have semantic interpretations, what we may regard as pure language abstracted from the sensory-motor systems used for externalization, which are unrelated to language. Of course, the kinds of computations that are ignored are readily within normal computational capacities, but they are not activated for language.

To take a very simple example, in "[the bombing of the cities] *copula* a crime," the copula is singular *is*, not plural *are*.[8] Instead of using the computationally trivial property of adjacency, the internal language relies on the abstract mental structure indicated by brackets, and the nontrivial operation of locating the *head* of the construction, which determines its syntactic/semantic role. To take another case, in the sentence "can [eagles that fly] swim," we do not associate the modal *can* with the linearly closest verb *fly*, a simple computation on what we actually hear (linear order), but with *swim*, the structurally closest verb, as the brackets indicate. The same holds for rules for all constructions in all languages.

Studies of language acquisition show that these properties are understood as early as relevant testing is possible. They show that reflexively, without experience, the infant ignores 100 percent of what it hears and avoids simple computations on these data, but rather attends only to what its mind creates, adopting without evidence the principle of structure dependence.

Your experiments with a variety of materials based on linear order showed that this curious property of our mental life is exhibited in the operations of the brain. That provides neural grounding to the distinction between possible and impossible languages discussed more extensively in your book *Impossible Languages.* In my judgment at least, these are the most revealing insights of neurolinguistics to date.

It's worth noting that there has been a great deal of work seeking to show that it would be possible to learn the principle of structure dependence by massive data analysis, or that hierarchical structures are found elsewhere in other mental acts or in nature. The proposals collapse on examination, but more to the point, they are irrelevant. They avoid the crucial question: Why, from infancy, do we ignore 100 percent of the data available to us as well as simple computations easily within the cognitive repertoire, instead attending only to what our minds create and we never hear? In the terms of your experiments, why does the brain not activate the canonical circuits for language when facing impossible languages with properties detectable by simple computations?

It should be added that by now there is a solid explanation for these discoveries about language. Structure dependence follows at once from the null hypothesis: once the principle of recursive generation of a discrete infinity of objects became available in evolutionary history, very likely along with *Homo sapiens*, nature as usual selected the simplest possible such procedure. There is, accordingly, no learning.

This array of empirical and conceptual arguments does not, of course, definitively prove that linear ordering appears nowhere in the workings of the internal language. That would be impossible to establish. But it does mean that claims that ordering does appear face a heavy burden of proof, empirically but also conceptually: to explain how and why such a serious departure from the optimal process could have taken place.

Questions of this kind have rarely arisen in linguistic inquiry, or in the cognitive sciences in general. I think we have, however, reached the point where they are becoming appropriate.

AM Nowadays, as we've discussed many times together, the new challenge is to shift from what I used to call the "where" problem—that is, *where* in the brain a certain network is active, say, the circuit that pertains to language as opposed to other cognitive capacities—to the "what" problem—that is, *what* is the actual information one neuron passes to another. But again, without a formal and explicit—in short, without a "generative" linguistic theory as a background, you cannot even start thinking about this. In this case, the famous, ubiquitous expression "big data" doesn't really count. Capturing the syntax of human languages in full by analyzing an immense number of sentences would be like capturing the fact that the sun is fixed and we rotate around it by taking thousands of trillions of pictures of the sun through the window. Scientific research simply doesn't develop like that, although in principle it could; moreover, even if big data research could give us an approximation of language structure, the probability that statistics could fully capture the actual mechanisms of the brain, as opposed to simply simulate them, would be very low, let alone that they could mimic the kinds of errors children make when they acquire their own grammar. All in all, in my own limited experience, I think that the experiments that can be conceived in the field, especially those involving syntax—the core of human language—can only be done if we take the generative procedures and the corresponding explanatory force you had envisaged and designed back in the '50s as a guideline.

But following up on your reflection on the danger of a new kind of reductionism, I would like you to share with us your thoughts on a related observation you once made. I remember you talked in class about the notion of gravity. During the Cartesian era, orthodox philosophers thought that, since no one would reasonably assume action at a distance, the trajectory of the Moon was explained by assuming that it is trapped in an etheric vortex centered on the Earth and so rotates around it; which is to say, it was thought to be fully mediated through a chain of direct local contacts. But then Newton's conclusions on gravity started circulating, which offered a radically different approach, provisionally adopting action at distance for descriptive reasons. I would like to ask you to expand on your own reflections on this specific historical circumstance and set of ideas. Because to me, they do outline the core challenge that neuroscience is facing today, similar to the one gravity posed in those times.

NC Well, a little bit of background, from earlier centuries. As I mentioned, there's a certain similarity between the kind of euphoria that Bar-Hillel was describing and the "big data" euphoria today: the sense that we have an answer to everything. Not for the first time. Something similar was true of the sixteenth century, the neo-scholastic period. Neo-scholastic physics had a lot of results; it described a lot of things very well. And there were apparent answers to just about everything. Suppose I'm holding a cup in my hand, and it has boiling water in it, and I let it go. The cup falls to the ground, the steam rises to the sky. Why? They had what passed for an answer. They're moving to their natural place. The cup is moving to the Earth, which is the natural place for solid objects; steam is going to the heavens, the natural place for gas. If two objects attract and repel each other, that's because they have sympathies and antipathies. If you look at the figure of a triangle and you see a triangle, the reason is that the form of the triangle moves through the air, gets into your eye, and implants itself in the brain. So, you have an answer to that problem. And, in fact, there were kinds of answers to almost everything. Much like the structuralist period, the Markovian information theoretic period, and so on.

Galileo and his contemporaries achieved something extremely significant. They allowed themselves to be puzzled. They thought: "Wait a minute." These descriptions are based on what they came to call "occult ideas": ideas that really have no substance. Like the idea that was near orthodoxy back in the '40s and '50s that "language is a matter of training and habit," the formulation of Leonard Bloomfield, the great American linguist. Others believed the same thing: Children are trained with thousands of examples, millions of examples, and somehow the habit is formed and they know what to say next. If they produce or understand something new, it's by "analogy." When you start thinking about it, it's completely inconceivable, to the extent that there's any substance at all.

Well, Galileo and his contemporaries took the same position toward neo-scholastic science. They said, none of this makes any sense. And they started to do what in fact were mostly thought experiments. Galileo never carried out most of his experiments. And if he had, they wouldn't have worked, because the equipment was too primitive. He worked out what "must" happen. So, he didn't drop balls off the Tower of Pisa. He developed very intelligent thought experiments, which showed that if you had a big heavy lead ball and a small lead ball, they'd have to fall at the same speed. A good argument.

This didn't cut much ice with the funders of the time, the aristocrats, the "national science foundation" of the day. They couldn't understand why anyone would study a ball rolling down a frictionless plane, something that doesn't even exist, when he could be studying something interesting, like the growth of flowers, or the sunset, or something like that. So, it was a hard job to try to convince them that, look, it's worth understanding these very simple things. If you have a ball on the top of a mast of a sailboat, and the sailboat is moving along, why does the ball fall to the bottom of the mast, why doesn't it fall behind, because the sailboat is moving forward? Notice that's an experiment that you could never carry out, because if you tried to carry it out, the ball would fall all over the place. But he showed just by thought experiments that, yes, there is a reason why it is going to fall to the base. And that's how modern science developed.

But it developed in an interesting way. The new scientists, Galileo and the rest, wanted to have a *serious* explanation. And they came up with what was called the mechanical philosophy. Now this was in part stimulated by something that was happening in Europe at the time. Skilled artisans at the time were creating very complex artifacts—complex clocks that did all sorts of things; constructions acting out plays with figures that were artificial but looked almost real; the gardens at Versailles, where you walk through the gardens and objects were carrying out all sorts of actions; and so on. Europe was full of such complex artifacts. Later a model of a duck digesting, one of Jacques de Vaucanson's models. All of this suggested that maybe the world was just a big example of a machine.

Just as an artisan can construct these incredible machines, which fool us into believing that they are alive, so a master artisan created the entire world as a super-complex machine.

Now, I think that there's an open question that could be studied today. My guess is that this is intuitive—our intuitive innate understanding of the world. So, for example, there was a famous experiment by Michotte back, I think, in the 1940s, where he showed that if you present a child with two bars, which are not quite touching, and one of them moves and then the other moves, the child will automatically assume that there is a hidden connection between them. In general, the mind kind of creates a mechanical explanation for whatever it sees happening, intuitively. And I suspect that investigation might demonstrate that the mechanical philosophy, as it was called, is just our intuitive sense of the world, fortified by what was happening with the development of complex artifacts. "Philosophy," of course, meant "science" at the time. It was mechanical science.

This was picked up by Descartes, as you mentioned, a great scientist, who thought he could demonstrate that the world was indeed a machine. Interestingly, he found one aspect of the world that wouldn't work like a machine: language. He said it's impossible to construct a machine that will produce expressions, the way we normally do, that are appropriate to situations but are not caused by them; as the Cartesians put it, we are incited and inclined to speak the way we do, but not compelled. We can act creatively and in our normal behavior create new thoughts, new expressions, that others can understand. In part to accommodate these facts of nature, Descartes postulated a new principle—in his metaphysics, a new substance, *res cogitans*, a thinking substance that underlies the normal use of language to construct thought.

Actually, Galileo and the Port Royal linguists, logicians, noticed the same thing in a different way. They expressed their awe and amazement at the fact that—it seemed miraculous, and in some ways still does—that with just a few symbols you can construct infinitely many thoughts, and you can convey to others, who have no access to your mind, its innermost workings. How is this miracle possible? A core problem of the study of language.

So, Descartes, Galileo, Arnauld, and others recognized that language and thought do not fall within the mechanical philosophy, but the rest of the world is a machine. Then Isaac

Newton came along. He was puzzled by Descartes's vortex theory of how things interact. The second volume of *Principia* is devoted to demonstrating that it doesn't work. So, what are we left with? Things attract each other, and repel each other; but there's no contact. Newton regarded this as what he called an "absurdity," that no person with any scientific knowledge could contemplate. The other great scientists of the day just dismissed it. Leibniz said, this is ridiculous: how could this conceivably be the case? Christiaan Huygens, the great experimentalist, said it was nonsense. It's reinventing occult ideas. And Newton actually agreed. He said, yes, it's like the occult ideas, but there's one difference. I have a theory that explains things, using these ideas. Now, he never called his work a philosophy of physics, or anything like that. Philosophy meant science. He just called it mathematical, a mathematical theory. The reason was, as he put it, "I have no explanations." His famous comment, "I make no hypotheses," was in that context. He said: "I have no physical explanation. I'm not going to make a hypothesis." And that's where it was left with Newton. In fact, a mechanical model provided the criterion of intelligibility for Galileo, Leibniz, Newton, and other great founders of modern science. If you didn't have a mechanical model, it wasn't intelligible. So, Galileo was dissatisfied with any theory of the tides because you couldn't construct a mechanical model for it.

Well, what happened after Newton is sort of interesting. Science just abandoned the hope for an intelligible world. Theories are intelligible, but what they describe is not intelligible. So, Newton's theory was intelligible. Leibniz did understand *that*. He couldn't understand what it described. That was unintelligible. It took a long time, but post-Newton science just slowly abandoned the search for an intelligible world. The world is whatever it is. The most we can hope for is intelligible theories. It took on a new, different form with Kant and with others. But essentially, science lowered its goals. So, if we can get an intelligible theory, then that's as far as you can go. We're not going to try to penetrate further. The goals of the great founders of modern science were abandoned.

It did take time. So, for example, in Cambridge, Newton's university, after his death—I think it was about half a century before they even started teaching his theories. Because they weren't real science, they were just mathematical accounts. Now this goes on into the twentieth century, in interesting ways. Take chemistry and physics. A century ago, chemistry was not reducible to physics. It was considered just a mode of calculating the results of experiments. Into the 1920s, Nobel laureates in physics and chemistry were describing chemistry as a mode of calculation. It's not a real science, because you can't reduce it to physics. Take Bertrand Russell, who knew the sciences very well. In 1928, he wrote that chemistry had not *yet* been reduced to physics. Maybe it will be someday, but we haven't gotten that far yet. Kind of like when people say today: Mental processes are not *yet* reduced to neural processes. But we'll get there.

Well, what happened with chemistry and physics? It turned out that chemistry could not be reduced to physics, because while chemistry was basically right, physics was wrong. Scientists came along with a new physics, and then you could *unify* a pretty much unchanged chemistry with physics, with the new physics. Directly, the quantum theoretic account of chemical properties. Linus Pauling gave a quantum theoretical account of the chemical bond and then he had a unified system—but no reduction. In fact, chemistry is not reducible to the physics of a century ago.

Well, let's come to today. The neurosciences have made progress, but they're nowhere near as advanced as physics was in the 1920s. That's not a criticism. It's much too complex. It might very well turn out that looking for reduction is the wrong way to proceed, that just as you couldn't reduce the world to mechanical models, and you couldn't reduce chemistry to physics, because the base for the reduction was wrong, it might turn out that it's the neurosciences, the brain sciences, that will have to be reconstructed in different ways if we're going to be able to unify them with what we discover about the nature of language, thought, cognition, and so on.

I think that there is some indication that that might be true. I'm thinking in particular of Randy Gallistel's work. A friend of ours who's a great cognitive neuroscientist and has been arguing for some years, with increasing resonance in the field, that neural nets are simply the wrong place to look for neural computation. Actually, Helmholtz, back in the nineteenth century, already had some reason to believe this. Neural nets are slow. Neural transmission is of course fast by my standards, but slow by the standards of what the brain has to accomplish. And more importantly, as Gallistel has demonstrated pretty well, I think, with neural nets you just cannot construct the minimal computational element that's needed for the basic theory of computation: Turing computability. Basically, the way your computer works: the basic unit that's involved in computation can't be constructed out of neural nets; so, there must be something else. Probably at the cellular level, where there's vastly more computing power, maybe internal to the cell, maybe microtubules or something else.

And incidentally, neural nets are the basis for the deep learning systems. They're modeled on neural nets. They're perhaps just looking in the wrong place, which is why everything is done basically by brute force, very rapid analyses of huge amounts of data to discern regularities and patterns. That can be very useful, but, keeping to the language case, what is being learned? Take your possible and impossible languages. The systems work just as well for both cases, which means that they are telling us nothing about language. A measure of the empirical content of a theory is what it excludes, as Karl Popper observed.

So, you get things that *look* exciting, just as the artifacts that were constructed by artisans in the sixteenth and seventeenth century were exciting, but they are not providing a model for understanding how the world works. That's very likely to be true, in my opinion, in the areas we're discussing. Lack of reduction of mental life to the neurosciences of today may be because the base system hasn't been developed properly. When it is, we may find real unification.

AM This thought-provoking synthesis offers a lot to think about regarding language structure and human nature in general, and calls for careful studies and confrontations with scientists of all disciplines and philosophers that should characterize the debate of the coming years. For now, I would just like to add two observations and one question. The two observations are very circumscribed. I remember their impact on me in defining scientific method when you presented them in class.

The first one concerns a thought that you just mentioned without emphasizing it. That is something you wrote, back when you gave the "Managua Lectures," and actually you rephrased in a very articulated fashion in the talk you gave at the Vatican.[9] It is an expression of yours that has represented a fundamental turning point in my own personal life, but—I am sure—for all the students who heard it. You once said: "It is important to learn to be surprised by simple facts." Considering it carefully and analyzing it word by word, this sentence contains at least four different foci, so to speak: first, it makes note of the importance of the thought expressed ("it is important"); second, it refers to a learning process, an effort rather than to a personal inherited talent ("to learn"), and by doing so it emphasizes the importance of the responsibility to teach; third, it refers to the sense of wonder and curiosity as the very engine of discovery, and to an awareness of the complexity of the world that is, an observation that goes back to Plato and the origin of philosophy ("to be surprised"); finally, fourth, arguably the most striking and innovative observation, it states that simple facts make a difference ("by simple facts"). The sudden awareness of something that calls for an explanation, once the fog of habit has lifted, seems to be the real stuff revolutions' sparkles are made of: from Newton's legendary falling apple to

Einstein's elevator, from Planck's black body problem to Mendel's pea plants, the real force comes from asking questions about what all of a sudden doesn't seem to be obvious. Of course, it could be that one is exposed to a certain fact by chance, but, as Pasteur once put it, "In the fields of observation chance favors only the prepared mind," and this is why we need to learn how to be surprised.

Actually, certain simple facts can be visible to the mind's eye rather than to our direct vision. Owen Gingerich once made me realize how Galileo reached the conclusion that all bodies fall to the Earth at the same speed even if they have different weight, besides the obvious restrictions due to their shape: Galileo never amused himself by throwing objects from the Tower of Pisa. Instead, he reflected that if a heavy object fell faster than a light one, then when the two objects are tied together we would face a paradox: the lighter object should slow down the heavier one, but together they should fall faster since their total weight is greater than that of the heavier object on its own. Galileo, surprised by this simple mental fact, came to the fundamental conclusion that the only possibility is that these two objects had to fall at the same speed and then, generalizing it, that all objects fall at the same speed (disregarding friction with the air due to their shape). And this without having to climb the tower other than to enjoy the panorama.

And the second thing I would like to highlight from your synthesis: at a certain point you said that it is impossible to build a machine that talks. Obviously, I cannot but agree, but there's one important thing that I want to emphasize: there is a fundamental distinction between *simulating* and *comprehending* the functioning (of a brain but also of any other organ or capacity). It is, of course, very useful to have tools, which we can interact with by "speaking," but it is certainly clear that those simulations cannot be used to understand what really goes on in the brain of a child when they grow and acquire their grammar. Of course, we can always stretch words so that they become felicitous to mean something different from what they used to mean. This reminds me of the answer Alan Turing gave to those who repeatedly asked him if one day machines could think. We can read his own words and substitute *think* with *talk*, which I think leaves the essence of Turing's idea valid: "I propose to consider the question, 'Can machines think?' This should begin with definitions of the meaning of the terms 'machine' and 'think.' The definitions might be framed so as to reflect so far as possible the normal use of the words, but this attitude is dangerous. If the meaning of the words 'machine' and 'think' are to be found by examining how they are commonly used it is difficult to escape the conclusion that the meaning and the answer to the question, 'Can machines think?' is to be sought in a statistical survey such as a Gallup poll. But this is absurd. . . . The original question, 'Can machines think?' I believe to be too meaningless to deserve discussion. Nevertheless I believe that at the end of the century the use of words and general educated opinion will have altered so much that one will be able to speak of machines thinking without expecting to be contradicted."[10]

But besides these two observations, there is one question I would like to ask you concerning these ideas. The way that you have depicted the relationship between chemistry and physics in the history of science allows us to reflect on the relationship between linguistics and neuroscience. My personal view, which doesn't count, obviously [*laughs*], and which is why I want to ask you, is that linguistics cannot be, *must* not be ancillary to what we currently know about our brain; but if anything, we have to change and grow toward, perhaps, a unification—*provided* that we dare to use the term "mystery" in the way that you used it. In other words, it is not out of the question that humans may never end up understanding creativity in language, namely the capacity to express a verbal thought independently of one's physical environment, Indeed, it could well be that we must just stop short of "the boundaries of Babel," that is, the limits of variation that may affect human languages as given independently of experience.

Equivalently, one could consider the boundaries of Babel as the infants' "stem mind," or "stem brain," that is, the potentiality to acquire *any* language within a certain amount of time since the birth. The discovery of this amazing link between language *structure* and the brain is so revolutionary that it can be expressed by reversing the two-thousand-year-old traditional perspective and arriving at the surprising conclusion that it's flesh that became *logos*, not vice versa.[11] I would like you to comment a little on this.

NC I'm kind of a minority. The two of us are a minority. [*Moro laughs.*] There may indeed be a mystery. Let's take a look at, say, rats, or some other organism. You can train a rat to run pretty complicated mazes. You're never going to train a rat to run a prime number maze—a maze that says, "turn right at every prime number." The reason is, the rat just doesn't have that concept. And there's no way to give it that concept. It's out of the conceptual range of the rat. That's true of every organism. Why shouldn't it be true of us? I mean, are we some kind of angels? Why shouldn't we have the same basic nature as other organisms? In fact, it's very hard to think how we cannot be like them. Take our physical capacities. I mean, take our capacity to run 100 meters. We have that capacity because we *cannot* fly. The ability to do something entails the lack of ability to do something else. I mean, we have the ability because we are somehow constructed so that we can do it. But that same design that's enabling us to do one thing is preventing us from doing something else. That's true of every domain of existence. Why shouldn't it be true of cognition? We're capable of developing—humans, not me—humans are capable of developing, say, advanced quantum theory, based on certain properties of their mind, and those very same properties may be preventing them from doing something else. In fact, I think we have examples of this; plausible examples. Take the crucial moment in science when scientists abandoned the hope for getting to an intelligible world. That was discussed at the time. David Hume, a great philosopher, in his *History of England*—he wrote a huge history of England—there's a chapter devoted to Isaac Newton, a full chapter. He describes Newton as, you know, the greatest mind that ever existed, and so on and so forth. He said Newton's great achievement was to draw the veil away from some of the mysteries of nature—namely, his theory of universal gravitation and so on—but to leave other mysteries hidden in ways we will never understand. Referring to: What's the world like? We'll never understand it. He left that as a permanent mystery. Well, as far as we know, he was right.

And there are other perhaps permanent mysteries. So, for example, Descartes, and others, when they were considering that mind is separate from body—notice that *that* theory fell apart because the theory of *body* was wrong; but the theory of mind may well have been right. But one of the things that they were concerned with was voluntary action. You decide to lift your finger. Nobody knows how that is possible; to this day we haven't a clue. The scientists who work on voluntary motion—one of them is Emilio Bizzi, he's one of MIT's great scientists, one of the leading scientists who works on voluntary motion—he and his associate Robert

Ajemian recently wrote a state-of-the-art article for the journal of the American Academy of Arts and Sciences in which they describe what has been discovered about voluntary motion. They say they'll put the outcome "fancifully." It's as if we're coming to understand the puppet and the strings, but we know nothing about the puppeteer. That remains as much a mystery as it has been since classical Greece. Not an inch of progress; nothing. Well, maybe that's another permanent mystery.

There are a lot of arguments saying, "Oh, it can't be true. Everything's deterministic," and so on. All sorts of claims. Nobody truly believes it, including those who present reasons (two thermostats might be hooked up to interact, but they don't take the trouble to work out reasons). Science doesn't tell us anything about it. Science tells us it doesn't fall within science, as currently understood. Science deals with things that are determined or random. That was understood in the seventeenth century. It's still true today. You have a science of events that are random, of things that are determined; you have no science of voluntary action. Just as you have no science of the creativity of language. Similar thing. Are they permanent mysteries? Could be. Could be that it's just something that we'll never comprehend.

Something similar might hold for some aspects of consciousness. What does it mean for me to look at the background that I see here and see something red? What's my feeling of red? You can describe what the sensory organs are doing, what's going on in the brain, but it doesn't capture the essence of seeing something red. Will we ever capture it? Maybe not. It's just something that's beyond our cognitive capacities. But that shouldn't really surprise us; we are organic creatures. It's a possibility.

So maybe the best that we can do is what science did after Newton: construct intelligible theories. Try to construct the best theory we can about consciousness or voluntary action or the creative use of language, or whatever we're talking about. The miracle that so amazed Galileo and Arnauld—and still amazes me, I can't understand it—how can we, with a few symbols, convey to others the inner workings of our mind? That's something to really be surprised about, and puzzled by. And we have some grasp of it, but not a lot.

When I started working on the history of linguistics—which had been totally forgotten; nobody knew about it—I discovered all sorts of things. One of the things I came across was

Wilhelm von Humboldt's very interesting work. One part of it that has since become famous is his statement that language "makes infinite use of finite means." It's often thought that we have answered that question with Turing computability and generative grammar, but we haven't. He was talking about infinite *use*, not the generative capacity. Yes, we can understand the generation of the expressions that we use, but we don't understand how we use them. Why do we decide to say this and not something else? In our normal interactions, why do we convey the inner workings of our minds to others in a particular way? Nobody understands that. So, the infinite *use* of language remains a mystery, as it always has. Humboldt's aphorism is constantly quoted, but the depth of the problem it formulates is not always recognized.

AM At least as linguists we can go on doing what you once told us, to use a slogan by Jean Baptiste Perrin, a French Nobel laureate in physics, is the fundamental strategy of science: namely, to reduce what is visible and complex into what is invisible and simple. More specifically, in linguistics, we need to search for the primitive, finite set of discrete elements that enter into the computation and the basic operations that assemble them to generate the potentially infinite array of linguistic expressions. In the introduction to your milestone publication "Lectures on Government and Binding" in 1981, you suggested that the agenda of syntax should be similar to the one of phonology. The idea was to abandon the classic taxonomy of linguistic rules including, for example, question formation, passivization, etc., and see them as the result of an interaction of more abstract entities. For example, language sounds are bundles of more abstract features—which is to say, the difference between the sounds represented by "f" and "v" in *fine* and *vine* is partially similar to the difference between the sounds of "s" and "z" in *see* and *zee*, traditionally named as "unvoiced" versus "voiced" sounds. So, the sounds "f" and "s" are not monolithic objects but the result of the assemblage of smaller shared features, and, in principle, all sounds of language are made by the assemblage of opposite features like this one. With syntax, this "minimalist search," as you dubbed it back in the mid-1990s, is in fact much more complicated and ambitious: finding the primitive elements and the combinatorial operation of syntactic rules or principles is obviously more difficult than with sounds, since we cannot even rely on physical or articulatory facts (and in fact the only undisputed physical and articulatory fact concerning language, namely linearity of the signal, is totally irrelevant for syntax). The core of this revolutionary approach to syntax is elegantly synthesized in your quotation: "These 'rules' (passive, relativization, question formation, etc.) are decomposed into the more fundamental elements of the subsystems of rules and principles. This development represents a substantial break from earlier generative grammar or from the traditional grammar on which it was in part modelled. It is reminiscent of the move from phonemes to features in the phonology of the Prague school."[12]

You often describe this crucial change of paradigm as coming from two basic types of facts: the substantial similarity of syntactic restrictions across languages, despite their apparent diversity, and the substantial similarity of the process of language acquisition by children in all languages, despite the subjective perception of the alleged different complexity between languages adults may develop. The only solution that can hope to capture these two facts in a unified way was the one you proposed: first, there is a unique set of primitive elements (beside Saussurian arbitrariness) and simple basic combinatorial operations; second, the interaction of these primitive elements and the principles generate a very complex system that allows certain restricted degrees of freedom, possibly solely linked to the primitive elements themselves, technically referred to as "parameters." The astonishing variety we perceive as adults between languages is the effect of small variations in such a unique complex system, but it should not be completely surprising, insofar language is considered from a biological perspective. In fact, it should rather be reminiscent of the one observed across the domains of living organisms, where the different species of animals, for example, are the result of the diverse order and quantity of the few same "letters" in the sequence of the same polymer molecule (DNA). All children are born equipped with the same basic set of primitive elements and operations which could potentially yield any language; witness the fact that the parents' language does not influence the child's if the child is raised in a community that speaks a different language. This is the formal characterization of the infant's "stem mind" when it comes to language acquisition and reduces it to the combination of parameters selected by the environment that becomes fixed as the child's definitive grammar. In other words, generative grammar could be considered as the limit of the influence of experience on language structure.

Along with this type of evidence, to support this "Mendelevian view" of language, I could perhaps add my own experience in the experimental field. Without this new approach to linguistics, there would have been no hope of exploring the neurobiological correlates of syntax and then arriving at the characterization of the "stem mind." The traditional taxonomy is far too removed from what we know of the brain's actual mechanisms to be used as a guide to inspect actual neurobiological networks. On the other hand, we have not reached a satisfactory level of abstraction (yet), but this is the same situation faced by every empirical science, prototypically in physics. Nevertheless, some crucial characteristics of these operations now seem to be reasonably identified, such as their ability to be applied an indefinite number of times (what we refer to as, in technical terms, "recursion"), their "blind," that is, nonteleological, character, or the role of instability as the core enzyme for the blossoming of those "snowflakes of words" we call sentences, to employ a beautiful metaphor you used to describe language as a whole. In particular the role of "unstable" structures in the minimalist project as linked to the notion of symmetry reminds me of Alan Turing's own words. In a different context, while trying to reduce the varieties of morphogenesis in biology to the interaction of few simple elements, he considered instability as a core engine: "It is suggested that a system . . . although it may originally be quite homogeneous, may later develop a pattern or structure due to an instability of the homogeneous equilibrium, which is triggered off by random disturbances. . . . It is found that there are six essentially different forms which this may take."[13] Similarly, some fundamental aspects of the variety of syntactic structures may be the result of a relative instability caused by the generation of few basic patterns involving symmetry.[14]

This fascinating dissection of human languages sometimes reminds me of a sensation that I experienced in my childhood, when, defying all prohibitions, I secretly looked at the hidden side of tapestries in museums. It pleased me to realize that the patches of colour that appear on the visible side, forming such elaborate pictures, were nothing but threads sewn in and out of the cloth in such a way as to form unexpected connections between different parts of the design. Unfolding the arborescent syntactic structures compressed into the deceivingly simpler linear sequences of words, observing their hidden warp and weft, gives me the same sensation and leads me to the conclusion that only what plays a role in an explanation can be said to exist.

My question now is: what do you anticipate the next steps will be toward this search for the minimal components of language and for a better perspective on this hidden side of language?

NC Predicting the future of scientific inquiry is venturing into uncharted waters, perhaps even a fool's errand. Maybe some great new ideas will come along, surprising us all. But keeping to what we now more or less understand, some challenging possibilities come to mind.

One is extending the neurolinguistic paradigm you developed to new areas. Structure-dependence, a fundamental property with far-reaching consequences, is well established on conceptual and empirical grounds, now extended to neurolinguistics. There are other claimants to a similar status. You brought up locality constraints, found in many areas. Presumably they have a common origin, perhaps in some notion of minimal search. Looking further, restriction of search holds outside of locality, as recent investigations have shown. And restriction of search in the domain of displacement you considered, the major area of inquiry into its application, raises the question why displacement takes place at all—not just displacement but also automatic "reconstruction," the fact that in such sentences as "what did John see" we understand "what" as the object of "see." Are there neural correlates that could shed some light on these very live topics?

As soon as the questions are raised, numerous others immediately come to mind. For example, could neurolinguistics inquiry shed light on the many questions that arise about in situ constructions with similar interpretations and properties but no displacement (or partial displacement), or about the residues left by successive-cyclic displacement?

The deeper question of why displacement takes place brings up the topic of dynamic antisymmetry that you've explored in depth, in particular, as a factor motivating displacement. Here too many questions arise. Why are such structures "unstable," requiring displacement? How does this factor in displacement interact with other considerations that have been proposed, among them Vergnaud's abstract case theory and identification of categories (labeling)? Could some of the many dilemmas that arise be illuminated by neurolinguistic inquiry?

Until recently, the ubiquitous property of displacement-reconstruction was regarded as a problematic complexity of natural language, contrasting with the expected operation of joining two elements together—often forming instability that has to be overcome by displacement. We now have good reason to believe that the opposite is true, for reasons related to the minimal search condition that figures prominently in locality. The considerations that arise are much like those you brought up in connection with parameters and Prague-distinctive features of phonology. And in science generally, as it pursues the objectives captured in Perrin's pithy phrase that you quoted. Why, for example, are there these chemical elements, not others, and what are the hidden components of which they are constituted?

Throughout, could neurolinguistic inquiries be designed to provide insights into the many puzzles that appear as soon as we try to move beyond description to genuine explanation?

These musings scarcely begin to trace even one strand of the intricacies of language. Puzzles and challenges abound, maybe even mysteries that are beyond our cognitive reach. We've traveled far from the days when there seemed to be answers to just about everything. A good sign.

The current situation reminds me of the title of a collection of essays in honor of one of the most outstanding contemporary philosophers, and close friend until his last days, Sidney Morgenbesser: *How Many Questions?*[15]

AM Our time is now up. Let me thank you for this extremely interesting conversation. I feel some nostalgia for the time when we could do this in person, but in the name of the organizers of the Festivaletteratura di Mantova and the public, let me thank you for all you've said and the time you've dedicated to us. This was a great conversation, for me, for us all. Thank you so much.

NC A pleasure: wonderful to talk to you.

Tucson and Pavia, July 2021

What Remains of the Future: Marginal Notes to a Conversation

It is an exercise in which many of us engage in many walks of life, a question to think through: What will remain of the present? What will remain of today's music in five hundred years? What will remain of today's literature? What will biology and physics look like at that point? Identifying with those who lived five hundred years ago and extrapolating from that is of little help, since history never allows for easy forecasts. Consider literature, for example: in 1521, one could be reasonably sure that Dante's *Comedy* (by then already "Divine" for two hundred years) would remain, but it was not obvious that Ariosto's *Orlando furioso*, for example, would withstand the oxidation of memory.

The same may be asked of the study of human language, and while the question may be futile, it is not foolish, as it compels us, even if only in our imagination, to dispense with the ideologies that can too often interfere with scientific judgment. Common belief may hold that the development of science is perfectly rational—mechanical and algorithmic, even—but this is far from the case. This is especially true when it comes to language, because language itself is something of a guiding thread that characterizes the thought of every epoch—a "Homeric question," as some scholars would put it. In fact, we could go so far as to say that the interpretation of human language is *the* Homeric question of humanity:

one that over time embodies the dominant outlook and reveals our essential traits, some of which we may not care to make explicit, either because we've overlooked them or because they appear to be so obvious that they don't need to be stated.

What, then, will remain of what we know of human language today? Let's take a small step back. In the 1950s, in the midst of the structuralist revolution that, starting from Ferdinand de Saussure in Geneva, had spread not only to the rest of Europe and the other continents but, even more importantly, to other domains besides that of linguistics,[1] only a few things about language were certain, but two of them were recognized as achievements. The first certainty was that Babel was a continent without boundaries: languages could vary "indefinitely and without limits" (as argued, authoritatively, by Martin Joos).[2] The second certainty was that the structure of this colossal artifact was a pure invention and the rules of languages "arbitrary, cultural conventions," comparable, as Eric Lenneberg (in his argument against this stance) put it, to the rules for card games, sports, or chess.[3] A century later, we still have few certainties about language, perhaps even fewer than we did then, but what we do know is that those two certainties—unlimited variation and the pure conventionality of language's rules—have turned out to be completely false: languages do not vary indefinitely but are instead circumscribed by robust formal constraints that limit the types of rules they follow,[4] and "the boundaries of Babel" are by no means "arbitrary, cultural conventions"; they are, rather, the expression of the neurobiological structure of the human brain.[5]

As in any scientific domain, a revolution is not the expression of a single individual but the reflection of the Zeitgeist that had allowed it to take root; the trigger, however, can often be attributed to the intuition of a single individual. In the case of the study of human language, it was Noam Chomsky's program of research, technically known as "generative grammar" (or "explicit grammar"), that spurred this radical change in perspective and demolished the two false beliefs that had been guiding linguists. But the driving force behind this change of paradigm is not that of simple demolition: to my mind, it can be summed up in three distinct, interdependent contributions.

Chomsky's first contribution was to ascribe a characteristic domain of grammar, syntax, to the broader scope of the empirical sciences: a domain of no small importance and one that would dominate linguistic research throughout the second half of the twentieth century. This focus on syntax overshadowed the prevalence of phonological and morphological analysis that had characterized the immediately preceding period, to such an extent that the one following it would be hailed as "the age of syntactic theories."[6] This epistemological and methodological restructuring can itself be broken down into three parts.

First, syntax lends itself to a relatively more immediate mathematical description than, for example, semantics (although the latter is certainly not exempt from it):[7] given a set of words as primitive elements, syntax is the function that generates all the possible sequences of words and therefore can be handled through an algebra or, in any case, through discrete mathematics. A more accurate way to describe it would be to assume that syntax generates an infinite number of structured and unordered expressions, which then become sequences of elements only when an externalization process takes place—which is to say, when the brain prepares a linguistic expression in accordance with the extralinguistic conditions imposed by the articulatory system (whether the sentence is uttered or remains part of one's inner speech activity). It is important to note that *generation* is mathematical in nature and that it must not be confused with a *process*: it is only the identification of an infinite number of objects. Production, on the other hand, is a process that must meet extralinguistic requirements. This crucial distinction has deep and wide-ranging consequences; it excludes, among other things, the possibility of comparing the motor planning activities related to the articulation that is part of the process with syntactic generation, and that the evolution of those motor activities might play a role in the evolution of language structure.[8]

Second, it should be noted that before Chomsky's inaugural works, syntax had hovered between Saussure's *langue* and *parole*, structuralism's two basic dichotomies of language: a general system of oppositions versus a set of individual acts of speech. This dichotomy Saussure established was something he himself wavered on, and it became ambiguous in subsequent structuralist analyses (particularly in the case of clause structure—the formal analysis of sentences—which Saussure and many other influential linguists such as Antoine Meillet regarded as an act of *parole*).[9] Chomsky instead provided a coherent notion of a system of rules, technically called *competence*—in opposition to *performance*, the actual use of language in concrete situations[10]—and he included syntax, and more specifically clause structures, within the former. Chomsky, in fact, has regarded this distinction between competence and performance as a modern revival of Aristotle's distinction between the possession of knowledge (generation) and the use of knowledge (production, perception, mental acts of all kinds)—with the fundamental difference that generation could now be explicitly described through mathematical algorithms.

Third, Chomsky's decision to adopt the so-called Galilean style of research practically nullified what had been a common claim, as often enthusiastic as it was groundless: that human languages could be investigated scientifically but that their regularities could be captured automatically, just by statistically recognizing a sufficiently large amount of data.[11] Moreover, this idea that the structure of languages could be entirely captured automatically by a statistical recognition based on Markovian chains was considered to be valid not only from a methodological perspective, but also on psychological grounds, in the sense that the same type of statistical recognition was considered to be the only means by which all infants spontaneously learn their mother tongue—as if the stimulus received contained all the instructions needed for infants to construe their grammar. Chomsky's reaction to this claim addressed both levels: he demonstrated that the adopted statistical models (Markovian chains) could not capture the regularities of natural languages,[12] and he argued that the stimulus available to infants was too poor to be the source of so complex a structure as a human grammar: something more than the combination of a *tabula rasa* and a statistical model was needed to explain spontaneous language acquisition in children.[13] Moreover, the claim that a linguistic phenomenon should not be questioned according to experimental criteria but should wear its analysis on its sleeve, so to speak, is unique among all empirical sciences. If this method were adopted in physics, for example, it would be the equivalent to claiming that analyzing a large enough number of photographs of the sky taken from the windowsill would be enough to produce a heliocentric theory of the solar system: it is certainly not impossible, but it is not how the actual history of that scientific enterprise unfolded, nor could it have in any reasonable amount of time. Scientific discovery instead had to proceed through insights, trial-and-error strategies, hypotheses, idealizations, and perhaps even through a certain irrational aesthetic taste for symmetry: theories had to be constructed that were not deduced automatically through purely statistical analysis, but rather progressively approximated a true interpretation of phenomena.

The linguist and the infant follow the same path when it comes to language, then, except that the linguist acts voluntarily whereas for the infant the discovery of grammar is something that "happens" instinctively and effortlessly—or, as Chomsky put it, children learn grammar the way they learn to walk or digest (and no child learns how to digest by means of imitation). More explicitly, the spontaneous process of language acquisition is an unconscious one, a selection that is made from the abundance of information with which an individual is biologically endowed prior to any experience:[14] a process of selection that turns out to correspond to a neurobiological process of synapse "pruning" in the intricate forest of the brain.[15] This revolutionary model is sometimes known as, to use Jacques Mehler's very vivid and famous expression, "learning by forgetting."[16] The goal is to capture the "zero state" of the human mind, prior to each linguistic experience, or what we might call the linguistic "stem mind": a totipotent state in the sense that it is still open, for a limited amount of time, to any possible language.

The inclusion of syntax in the empirical sciences had two more consequences, the scope of which has not yet been entirely calculated: a consequence for the *comparative* domain (in the sense this word has in comparative biology), and another for the technological one.

When it comes to the first, many experiments have confirmed a hypothesis that Descartes had already asserted at an intuitive level, namely, that syntax constitutes the fundamental distinction between the language of humans and that of all other living beings: we are the only ones capable of recombining a bounded set of discrete elements (broadly, words) to generate a potentially infinite set of expressions (broadly, sentences). This is particularly evident with regard to our closest cousins, primates,[17] but it is also true for other animals such as birds.[18] If the consequence of this in the field of biology is positive, in that it provides new data and perspectives that confirm what had been a powerful intuition, it is otherwise for the field of technology: we know that machines can only approximate correct syntax and that this approximation is only a simulation of human behavior, not a comprehension of the actual neurobiological mechanisms underlying the capacity that makes this behavior possible. Indeed, we must necessarily make a sharp distinction between *simulation* and *comprehension*; otherwise, a dangerously deceptive situation is created. Needless to say, simulation can in fact be very useful, but it answers different questions than those aimed at discovering and comprehending the actual mechanisms underlying the faculty of language. The domain of biology and that of technology must not be confused—be it in the object of study, in the method employed, or in the purpose of the research. Many of the expressions in vogue today—from *artificial intelligence* (which has replaced the obsolete *cybernetics*), to *digital natives*, to the misleading idea that languages are a *software* freely running on the *hardware* of the brain—are at best metaphors, more useful for those aiming to identify and stimulate a new class of consumers than for those interested in understanding natural phenomena. The potential nightmare of an artificially designed language imposed on humans, however, is fortunately averted by the biological nature of the actual mechanisms of linguistic structures, as we will see when we discuss Chomsky's third contribution: for now, at least, such violence is conceivable only in novels.[19]

Chomsky's second contribution was to invent a formal apparatus and a method without which it would be impossible to discover and recognize the essential unity underlying the apparently accidental varieties of language, and to do so by emulating the method adopted for phonology by the Prague school.[20]

This was carrying over what had already taken place in physics, which ever since Galileo has understood the book of nature to be written in numbers; in linguistics, the primary goal has been, and still is, to identify a basic alphabet for those chapters of that same book that deal with language. This is not at all a marginal aspect, but it is too often underestimated, and it has given contemporary linguistics a simplistic, even caricatural, appearance. But the only reason any area of experimental science is able to progress in any substantial way (which is to say, beyond subjective intuitions) is by means of a formal language, one that is capable of selecting the salient properties of primitive elements and of constructing a compositional algebra to guide empirical research. Physics, chemistry, biology, genomics: no domain is exempt from this effort—so much so that research in any field of science can be summed up by Jean Perrin's ecumenical motto, which states that the task of science is to "explain what is visible and complicated by means of what is simple and invisible."[21] It is by means of this formal apparatus that an enormous mass of syntactic phenomena has been brought to light that had never been described before, in Indo-European languages as well as many other families. This epochal progress has been concretely documented in such works as the monumental encyclopedia of syntax edited by Martin Everaert and Henk van Riemsdjik,[22] but also by a flourishing number of comparative investigations on increasingly "microscopic" levels of analysis, such as the splitting of inflectional morphemes into smaller units, which began with the so-called Split-Infl hypothesis,[23] and developed into a very productive line of research known as "linguistic cartography."[24] And it is on the basis of such a broad empirical recognition and a unifying formal apparatus that the great theoretical syntheses that led to the construction of a general model of human syntactic competence were born.[25] The most important result was the unpredictable and unprecedented discovery that every rule and construction in every language adheres to the principle of structure-dependency based on the hierarchical structures generated recursively, the most fundamental property of language, rather than the flat sequences of words creating a linear order, ironically the only undisputed property of human languages.[26]

An example of structure-dependency can be found the discovery of the so-called structure-preserving principles and, among these, "locality restrictions." This term refers to the limits at which syntactic items can act from a distance when calculated in a two-dimensional metrics generated by assembling syntactic units in recursive binary fashion.[27] A prototypical example is the well-known phenomenon of "syntactic movement" (also known as "discontinuous constituents"),[28] such as the distance at which an interrogative pronoun can act with respect to the verb governing it, as in the following contrast: *What do you think Peter would like to taste before accepting the cookies?* vs. **What do you think Peter would like to taste some cookies before accepting?* There is no logical reason as to why the second sentence is ungrammatical, no logical reason as to why *what* cannot be the complement of *accepting*, whereas it can be the complement of *taste*. This cannot be attributed to an inability of the memory to retain the beginning of the sentence, because *taste* could be even more distant from *accepting* in terms of the number of words separating it from *what* and the sentence would be still grammatical, as in the case of: *What does Lucy think that John knows that Peter would like to taste before accepting the cookies?* The reasons for these restrictions are purely formal and cannot be traced back to any other cognitive component. The comparative analysis of locality conditions not only revealed the substantial invariance of these conditions across the languages of the world, but also allowed linguists to hypothesize and identify points of minimal, ideally binary, variation within the grammatical systems. These points, technically defined as "parameters" and often informally referred to as "switches," are associated with those possible states of grammars that are not predictable and derivable by principles. The idea behind parameter theory is that these minimal and unpredictable points of variations have such intricate effects in the overall architecture of grammar that languages can end up appearing dramatically different, to such an extent that language variation had previously been considered a qualitative, irreducible, and unbounded phenomenon. The fact that minimal variations can produce dramatic differences in complex systems is another way of reinforcing the hypothesis that language structures share many properties with biological phenomena, similarly to the way different species can result from genetic differences on the molecular level.

But experience is far from being an irrelevant factor in Chomsky's view of language acquisition; the scaffolding of invariant principles and parameters of variation can in fact be regarded as the formal identification of a biologically determined grid designed to limit the effects of experience on language variation, with obvious far-ranging consequences for language acquisition and evolution. This formal apparatus is what led to the construction of innovative syntactic theories: one of the emblematic examples is the hypothesis that syntactic combination is always binary.[29] This single hypothesis has allowed for many analyses and theories that have no immediate precedent in linguistics, such as the derivation of phrase structure from a single axiom[30] or the unified theory of phrase structure and syntactic movement.[31]

But this formal system's usefulness does not come solely from its synthetic and deductive power. This formal language ultimately has a strong heuristic power as well: it uncovers the hidden organization of reality, highlights non-apparent similarities, and allows us to enter regions whose existence we were not even aware before. It is as if you could look at the back side of a tapestry and see how the distinct dots composing the image in front are not really disconnected: those dots are in fact little portions of a thread that emerge from an intricate system whose hidden path is inaccessible to immediate direct vision; we have to reconstruct this path by making deductions from what we see, but in the end, this new perspective on the "back side" confirms and explains everything. In a sense, it is its hidden side that is real when it comes to a tapestry. And this hidden side looks far from chaotic when you understand the principles underlying the paths of the threads. It is the same formal system that allowed the periodic table of elements to become the prototypical model of scientific research, and which has allowed linguists to dream of a "periodic table" of human languages. The fact that this linguistic table is not yet available and has not been completed does not mean that it does not exist; no scientific table has ever been filled in quickly. But this powerful vision of the world's languages as variations on a single theme, like the unification that took place in biology with the advent of molecular biology, has obviously made a huge impact in all sectors, including anthropology; for example, it dissolves the distinction of quality between languages and their alleged individual merits—a devious and dangerous residue of the notion of race, and one that has contributed to the delusional ideology of the "purity" of the Aryan race.[32] Despite all superficial and apparent differences, every human being in essence speaks the same language, the same way that every human being has in essence the same face: that humans at a certain level of abstraction have the same face is an intuition that painters like Albrecht Dürer already had back in the sixteenth century. Given this, if we want to argue that every human being has a different face, we should assume that every human being also speaks a different language, and therefore no manner of perceiving reality or thinking can be considered advantageous based on the use of any specific given language. The typical objection made by many adults that some languages are objectively more complex than others is as a whole disproved by, among other things, the fact that children learn *any* language on average in the *same* amount of time.[33] For a child, the back side of the tapestry should not be too difficult to grasp, regardless of how the image on the front varies. But the feeling that one's own language is the best seems to have always resisted any attempt to dismantle language in a rational way. Even Dante had already noticed this, although his observation has gone largely unnoticed to this day. In *De vulgari eloquentia*, his incomplete treatise in Latin on natural languages, he writes: "Pietramala is a very large city, the homeland of most of Adam's children. Whoever is so misguided as to believe that where he was born is the most beautiful place under the sun, will also believe that his own vulgar—that is, his mother tongue—is superior to all other vulgaris, and from this he deduces that his language was the one used by Adam" (Dante, *De vulgari eloquentia* VI). For Dante, the idea that better languages can exist was *obscenus*, which means both "repulsive" and "unfortunate" in Latin, but also "ridiculous" given that he chose Pietramala as an example, which was a tiny, practically unknown village, or perhaps just a castle, in the Apennines halfway between Bologna and Florence. But one cannot convince someone by means of rationality to stop being in love with someone else, and the "obscenity" of Pietramala that Dante described remains an open problem that linguistics has not been able to solve.

Chomsky's third contribution, though in a certain sense it has been an indirect one, was to open the way to examining the relationship between the formal structures of syntax and the neurobiological structures of the brain. The development of this relationship is not always linear and has in some ways been contradictory. Chomsky had immediately recognized a connection at the psychological level: first, in his famous controversy with B. F. Skinner, and then in the one with Jean Piaget.[34] It was already clear at this early stage that children cannot spontaneously learn the syntax of human languages by trial and error or by being taught it, especially since adults are unable to directly inspect the system of syntax themselves. Moreover, as already noted, every child acquires *any* grammar on average in the *same* period of time and regardless of the language spoken by their parents. For this to be so, the neurobiological contribution that precedes experience must be valid for any language. But despite such illuminating, though severely limited, clinical evidence as Eric Lenneberg's data on aphasic patients, Chomsky was, until the early years of the twenty-first century, openly skeptical about the possibility that formal linguistics could communicate with neuroscience in any substantial way, at least compared with the advances made in the understanding of syntax. However, the advent of neuroimaging techniques[35] and a generation of researchers who began to systematically expand the methods and results of generative grammar into extralinguistic domains have provided a new harvest of different data. Consequently, Chomsky's attitude has since changed and the neurobiological basis of syntax is now a focal issue.[36] It is in fact no longer possible to investigate the neurobiological basis of syntax without making reference to generative grammar and, vice versa, linguistics cannot ignore the results of neuroscience if it is to pursue the description of the structure of language.[37] This broadening of the empirical domain in linguistics made for two central turning points in the very development of the formal research program: first, the recognition that children spontaneously acquire language effortlessly, with no instructions, and only by forming a subset of the entire range of possible errors, despite the poverty of the stimulus; second, the incorporation of neurobiological data, whether from Lenneberg's clinical studies of aphasiological subjects or from subjects without impairments by means of neuroimaging technology.[38] This does not mean, of course, that some research projects in neurolinguistics modeled on generative grammar aren't guilty of striking simplifications: very few of them, for example, take into account the principles of locality and syntactic dependency and focus solely on structural composition and hierarchical syntax, which can sometimes produce very misleading conclusions. One such example is the idea that sequences of actions are governed by the same formal principles that govern the sequences of words.[39] But the critical point is far more profound and radical: it is now clear that research in this area can no longer be limited to an attempt to correlate linguistic phenomena with those networks that process them (the "where" problem), and it is the utilization of neuroimaging technology that has allowed this breakthrough. It has also crucially demonstrated not only that there are boundaries to Babel, but that these boundaries are present in the body itself; in fact, these boundaries turn out to be the expression of the neurobiological architecture of the brain, reversing the established two-thousand-year-old view by claiming instead that *flesh became logos*, language. More explicitly, neuroimaging techniques have demonstrated beyond any reasonable doubt the existence of "impossible languages"—grammatical systems that may be consistent, complete, and perhaps even simple in nature, but which do not meet the specific formal requirements of human language and, as such, are simply not recognized as linguistic data and are not processed by the natural networks of language. More precisely, it has been demonstrated that the syntax of all natural languages, surprisingly, ignores the linear order of words, which is the only incontrovertible fact and one based on physical reality about language. In fact, syntactic principles are not linear, but are instead based on the hierarchical structure that results from combinatorial binary recursive rules.[40] But this achievement is only the logical premise for the more complex task: to overcome the "where" problem and attempt to decipher the electrophysiological code with which neurons exchange and process linguistic information (the "what" problem);

this goal is particularly difficult, of course, since brain signals contain more than just syntactic information, and include sound representation in non-acoustic areas of the brain such as Broca's area and even during inner speech (or, in technical terms, endophasic activity).[41] The fact is that this leap from the "where problem" to the "what problem" is difficult not just for the obvious technical problems involved, which could potentially be solved in a reasonable amount of time.[42] This leap suggests theoretical difficulties that may be insurmountable. Even the desired solution to the so-called problem of the different "granularities" between the primitive elements of language and those of their neurobiological structure may not be sufficient to pave the way. This problem, raised by David Poeppel, is challenging and interesting: at a first approximation, the problem of granularity consists in the fact that the structural distance between two primitive elements of two different domains, say a word and a neuron, is incommensurable, and no experimental analogy can be established between them. But the real problem, which is beginning to emerge more and more clearly, is that not only is linguistics not yet ready for this sort of unification, but neuroscience may not be either. This scenario is delicate and murky, and far removed from the propaganda of so-called scientific content commericals, and Chomsky has given it special attention, comparing it to other crucial moments in the development of science: the one that took place in physics in Newton's time, for example, when an understanding of gravitational phenomena was established, or with the quantum interpretation of chemical phenomena that took place in the twentieth century.[43] When two disciplines meet, any change that could lead to unification will never be unilateral, and this certainly applies to formal linguistics and the neurobiology of the brain. Should neurolinguistics ever emerge as an independent discipline, it will do so only through the combination of a new linguistics and a new neuropsychology, not from the absorption of the former into the latter. In science, unilateral reductions—"epistemological annexation," one might say—either make no sense or are just propaganda.

The themes outlined here in a synthetic and cursory way by no means exhaust the domain of linguistics: language is a universe, and it is a universe in constant change. Entire continents have been excluded in the preceding, the boundaries of which within the regions of syntax are not stable: formal semantics, for example, is currently undergoing an extraordinary transformation along the exact same methodological coordinates Chomsky had established for syntax (after all, the term "syntax" is more than two thousand years old, whereas "semantics" was created in 1897 by Michel Bréal);[44] morphology is today being radically reconceptualized;[45] and the same can be said for phonology and pragmatics.[46] Moreover, the mechanisms underlying how language changes over time remain essentially beyond our understanding, not because there are language changes—if there were none, that would be surprising—but because we cannot predict what they will be. Languages are like glaciers: we walk upon them, they move, but we do not recognize this movement unless we realize that the shape of the panorama has globally changed. In synthesis: language changes themselves are almost unpredictable, especially changes in syntax.[47] If this shifting and still quite obscure domain of linguistics, in which culture and history play major roles, remains obscure, it has nonetheless been possible to draw a sharp and fundamental distinction between language *evolution* and language *change*: languages have not evolved, they have changed, like variations on a unique theme. To say that languages have evolved would be like saying that a daughter's face is more evolved than her mother's—a clearly inconsistent conclusion: this distinction is also one of the results of the new way to look at language originally introduced by Chomsky. Finally, if we are to avoid a simplistic and fundamentally false vision of contemporary linguistics and, ultimately, of language as a whole, we must assume that every type of empirical data is pertinent: from the historical and diachronic to the synchronic, from the sociolinguistic to the stylistic, from data deduced from corpora to literary data, from that related to acquisition to aphasiological data, from the genetic to the paleontological, from mathematical and logical data to the neurophysiological—and even physical data, if we include the work being done on the wave structure of the electrophysiological code of language.[48] This ability to include all types of empirical data within the domain of observation without ideological bias is also characteristic of the method inaugurated by Chomsky. But he adds caution to this open-mindedness, evoking Emil du Bois-Reymond's disarming but necessary motto, *ignorabimus*:[49] he explicitly admits that, compared with those mysteries one can reasonably expect to see resolved relatively quickly with the advance of technology, there may also remain mysteries forever inaccessible to humans: the *big bang* question of how language originated in our species, for example.[50]

A few remarks to conclude these notes on Noam Chomsky's impact on the theory of syntax in human languages. The inclusion of syntax among empirical sciences, the development of a formalism dedicated to syntax, and the measurement of the neural correlates of syntactic phenomena all converge on a twofold goal: the identification of "the boundaries of Babel" and their naturalization, which is to say, to understand the limits of syntactic variation as the expression of the neurobiological structure of the brain. I emphasize *the limits* because it is essential to note that it is only a matter of capturing, in a certain sense, the perimeter of these boundaries, not what they contain and what changes there will be: that is left to a level of analysis that includes history, culture, and ultimately creativity itself.

But metaphors aside, the task of contemporary linguistics is not so much to classify possible languages as it is to characterize what constitutes *impossible languages*, that is, to identify the limits of the variation of linguistic forms *within* a language and *across* languages—and this is where neurobiological experiments become substantial. This vision of human language, which in a sense was overturned when it shifted from possible to impossible languages and was anchored in the neurobiology of the brain, was unimaginable not only five hundred years ago, but even just fifty years ago: it is a vision that consists of a linguistic stem mind, with which all humans—and only humans—are endowed.

What will remain of contemporary linguistics in five hundred years? We may lack a clear answer, but if we are in any position to formulate novel questions for the future, the essence of science, we certainly owe it to Noam Chomsky's revolutionary vision of language.[51]

—*Andrea Moro*

Notes

The Secrets of Words

1. *Logic Methodology and Philosophy of Science: Proceedings of the 1964 International Congress*, ed.Yehoshua Bar-Hillel (Amsterdam: North-Holland, 1965).
2. Eric Lennenberg, *Biological Foundations of Language* (New York: Wiley, 1967).
3. Karl S. Lashley, "The Problem of Serial Order in Behavior," in *Cerebral Mechanisms in Behavior*, ed. L. A. Jeffress, 112–146 (New York: Wiley).
4. B. F. Skinner, *Verbal Behavior* (New York: Prentice Hall, 1957); Noam Chomsky, "A Review of B. F. Skinner's *Verbal Behavior*," *Language* 35 (1959): 26–57.
5. M. Musso, A. Moro, V. Glauche, M. Rijntjes, J. Reichenbach, C. Büchel, and C. Weiller, "Broca's Area and the Language Instinct," *Nature Neuroscience* 6 (2003): 774–781.
6. M. Tettamanti, H. Alkadhi, A. Moro, D. Perani, S. Kollias, and D. Weniger, "Neural Correlates for the Acquisition of Natural Language Syntax," *NeuroImage* 17 (2002): 700–709.

7. Lenneberg, *Biological Foundations of Language*, 2.
8. Andrea Moro, *The Raising of Predicates* (Cambridge: Cambridge University Press, 1997); Andrea Moro, *A Brief History of the Verb* To Be (Cambridge, MA: MIT Press, 2018).
9. Noam Chomsky, *Il mistero del linguaggio: Nuove prospettive* (Milan: Raffaello Cortina Editore, 2018).
10. Alan M. Turing, "Computing Machinery and Intelligence," *Mind* 59 (1950): 433–460, 442.
11. John 1:14. See Joseph Ratzinger, *Introduction to Christianity*, 2nd ed., trans. J. R. Foster (San Francisco: Ignatius Press, 2004), 176.
12. Noam Chomsky, *Lectures on Government and Binding* (Dordrecht: Foris, 1981), 7.
13. Alan Turing, "The Chemical Basis of Morphogenesis," *Philosophical Transactions of the Royal Society of London, Series G: Biological Sciences* 237, no. 641 (1952): 37–72.
14. Andrea Moro, *Dynamic Antisymmetry* (Cambridge, MA: MIT Press, 2000); Noam Chomsky, "Problems of Projection," *Lingua* 130 (2013): 33–49.
15. *How Many Questions? Essays in Honor of Sidney Morgenbesser*, ed. Leigh S. Cauman, Isaac Levi, Charles D. Parsons, and Robert Schwartz (Indianapolis: Hackett, 1983).

What Remains of the Future: Marginal Notes to a Conversation

1. Giulio C. Lepschy, *La linguistica del Novecento* (Bologna: Il Mulino, 2000), and Giorgio Graffi, *200 Years*

of *Syntax: A Critical Survey* (Amsterdam: John Benjamins, 2001).

2. Eric P. Hamp, Martin Joos, Fred W. Householder, and Robert Austerlitz, *Readings in Linguistics I and II* (Chicago: University of Chicago Press, 1995).
3. Eric Lenneberg, *Biological Foundations of Language* (New York: John Wiley, 1967).
4. Luigi Rizzi, "The Discovery of Language Invariance and Variation, and Its Relevance for the Cognitive Sciences," *Behavioral and Brain Sciences* 32 (2009): 467–468.
5. Andrea Moro, *The Boundaries of Babel: The Brain and the Enigma of Impossible Languages*, 2nd ed. (Cambridge, MA: MIT Press, 2015), and *Impossible Languages* (Cambridge, MA: MIT Press, 2016).
6. Graffi, *200 Years of Syntax*.
7. Barbara H. Partee, Alice ter Meulen, and Robert E. Wall, *Mathematical Models in Linguistics* (Dordrecht: Kluwer Academic, 1990); Denis Delfitto, "Linguistica chomskiana e significato: Valutazioni e prospettive," *Lingue e Linguaggio* 2 (2002): 197–236; Gennaro Chierchia, *Logic in Grammar: Polarity, Free Choice, and Intervention* (Oxford: Oxford University Press, 2013).
8. Andrea Moro, "On the Similarity between Syntax and Actions," *Trends in Cognitive Sciences* 18, no. 3 (2014): 109–110, Andrea Moro, "Response to Pulvermueller: The Syntax of Actions and Other Metaphors," *Trends in Cognitive Sciences* 18, no. 5 (2014): 221, contra Michael C. Corballis, *From Hand to Mouth: The Origins of Language* (Princeton: Princeton University Press, 2003).

9. Graffi, *200 Years of Syntax*.
10. Noam Chomsky, *Aspects of the Theory of Syntax* (Cambridge, MA: MIT Press, 1965).
11. Claude E. Shannon and Warren Weaver, *The Mathematical Theory of Communication* (Urbana: University of Illinois Press, 1949).
12. Noam Chomsky, "Three Models for the Description of Grammar," *IRE Transaction on Information Theory* IT-2 (1956): 113–124.
13. Noam Chomsky, "Review of Skinner 1957," *Language* 35 (1959): 26–58; Noam Chomsky, *Lectures on Government and Binding* (Berlin: De Gruyter, 1981); Massimo Piattelli-Palmarini, ed., *Language and Learning: The Debate between Jean Piaget and Noam Chomsky* (Cambridge, MA: Harvard University Press, 1980); Charles D. Yang, "Universal Grammar, Statistics or Both?" *Trends in Cognitive Science (TICS)* 8, no. 10 (2004): 451–456.
14. Massimo Piattelli-Palmarini, "Evolution, Selection and Cognition: From 'Learning' to Parameter Setting in Biology and the Study of Language," *Cognition* 31 (1989): 1–44.
15. Jean-Pierre Changeux, *L'homme neuronal* (Paris: Fayard, 1983), trans. Laurence Garey as *Neuronal Man: The Biology of Mind* (New York: Pantheon Books, 1985); Jean-Pierre Changeux, Philippe Courrége, and Antoine Danchin, "A Theory of the Epigenesis of Neuronal Networks by Selective Stabilization of Synapses," *Proceedings of the National Academy of Science PNAS* 70, no. 10 (1973): 2974–2978.
16. Jacques Mehler, "Connaître par désapprentissage," in *L'Unité de l'Homme 2: Le Cerveau Humain*, ed. Edgar

Morin and Massimo Piattelli-Palmarini (Paris: Éditions du Seuil, 1974), 25–37.

17. See, for example, Herbert S. Terrace, L. A. Petitto, R. J. Sanders, and Thomas G. Bever, "Can an Ape Create a Sentence?" *Science* 206, no. 4421 (1979): 891–902; Emilie Genty and Richard W. Byrne, "Why Do Gorillas Make Sequences of Gestures"? *Animal Cognition* 13 (2010): 287–301, https://doi.org/10.1007/s10071-009-0266-4; and Philippe Schlenker, Emmanuel Chemla, Anne M. Schel, James Fuller, Jean-Pierre Gautier, Jeremy Kuhn, Dunja Vaselinovič, Kate Arnold, Cristiane Cäsar, Sumir Keenan, Alban Lemasson, Karim Ouattara, Robin Ryder, and Klaus Zuberbühler, "Formal Monkey Linguistics," *Theoretical Linguistics* 42, nos. 1–2 (2016): 1–90.
18. Johan J. Bolhuis, Gabriël J. L. Beckers, Marinus A. C. Huybregts, Robert C. Berwick, and Martin B. H. Everaert, "Meaningful Syntactic Structure in Songbird Vocalizations?" *PLOS Biology* 16(6) (2018): e2005157; Gabriël J. L. Beckers, Johan J. Bolhuis, Kazuo Okanoya, and Robert C. Berwick, "Birdsong Neurolinguistics: Songbird Context-Free Grammar Claim Is Premature," *NeuroReport*, 23 (2012): 139–145.
19. Andrea Moro, *Il segreto di Pietramala* (Milan: La nave di Teseo, 2018); English translation (forthcoming) as *The Secret of Pietramala* (Milan: La Nave di Teseo).
20. See Chomsky, *Lectures on Government and Binding*, and Noam Chomsky, *The Minimalist Program* (Cambridge, MA: MIT Press, 1995); and, for a retrospective analysis, Noam Chomsky, *The Generative Enterprise Revisited* (Berlin: Mouton de Gruyter, 2004); Noam Chomsky, "Problems of Projection," *Lingua* 130 (June

2013): 33–49; Noam Chomsky, Ángel J. Gallego, and Dennis Ott, "Generative Grammar and the Faculty of Language: Insight, Questions and Challenges," *Catalan Journal of Linguistics, Special issue*, 2019; and Andrea Moro, *I Speak, Therefore I Am: Seventeen Thoughts about Language*, trans. Ian Roberts (New York: Columbia University Press, 2016).

21. Jean Perrin, *Les atoms* (Paris: Alcan, 1913), v.
22. Martin Everaert and Henk C. van Riemsdijk, eds., *The Wiley Blackwell Companion to Syntax*, 2nd ed. (London: Wiley Blackwell, 2017).
23. Inaugurated by Andrea Moro, "Per una teoria unificata delle frasi copulari," *Rivista di Grammatica Generativa*, 13 (1988): 81–110; Jean-Yves Pollock, "Verb Movement, UG, and the Structure of IP," *Linguistic Inquiry* 20 (1989): 365–424; and Adriana Belletti, *Generalized Verb Movement* (Turin: Rosenberg & Sellier, 1990).
24. Gugliemo Cinque and Luigi Rizzi, "The Cartography of Syntactic Structures," in *Oxford Handbook of Linguistic Analysis*, ed. Bernd Heine and Heiko Narrog (Oxford: Oxford University Press, 2009), 51–65.
25. Ian Roberts, "From Rules to Constraints," *Lingua e stile* 23 (1988): 445–464.
26. "Structure-dependency" is a compact way of saying that what matters in syntax is the hierarchical structure provided by simple binary recursive mechanisms of composition rather than the linear order (of words) expressed. This principle has eventually been supported by neurobiological evidence: see Moro, *Impossible Languages*.

27. Maria Rita Manzini, *Locality* (Cambridge, MA: MIT Press, 1992); Luigi Rizzi, "Labeling, Maximality and the Head-Phrase Distinction," *Linguistic Review* 33, no. 1 (2016): 103–127.
28. Kenneth L. Pike, "Taxemes and Immediate Constituents," *Language* 19 (1943): 65–82; Graffi, *200 Years of Syntax*.
29. Richard S. Kayne, *Connectedness and Binary Branching* (Dordrecht: Foris, 1984).
30. Richard S. Kayne, *The Antisymmetry of Syntax* (Cambridge, MA: MIT Press, 1994).
31. Andrea Moro, *Dynamic Antisymmetry* (Cambridge, MA: MIT Press, 2000).
32. Andrea Moro, *La razza e la lingua: Sei lezioni sul razzismo* (Milan: La nave di Teseo, 2019).
33. Frederick Newmeyer and Laurel B. Preston, eds., *Measuring Grammatical Complexity* (Oxford: Oxford University Press, 2014).
34. Chomsky, "Review of Skinner 1957"; Piattelli Palmarini, *Language and Learning*.
35. David Embick and David Poeppel, "Mapping Syntax Using Imaging: Prospects and Problems for the Study of Neurolinguistic Computation," in *Encyclopedia of Language and Linguistics*, 2nd ed., ed. Keith Brown (Amsterdam: Elsevier, 2005); Stefano F. Cappa, "Imaging Semantics and Syntax," *NeuroImage* 61, no. 2 (June 2012): 427–431.
36. Chomsky, *The Generative Enterprise Revisited*; Robert C. Berwick and Noam Chomsky, *Why Only Us: Language and Evolution* (Cambridge, MA: MIT Press, 2015); Angela Friederici, Noam Chomsky, Robert C.

Berwick, Andrea Moro, and Johan J. Bolhuis, "Language, Mind and Brain," *Nature Human Behavior* 1 (2017): 713–722, https://doi.org/10.1038/s41562-017-0184-4.

37. Everaert and van Riemsdijk, *The Wiley Blackwell Companion to Syntax*; Friederici et al., "Language, Mind and Brain."
38. Chomsky, *The Generative Enterprise Revisited.*
39. Andrea Moro, "On the Similarity between Syntax and Actions," *Trends in Cognitive Sciences* 18, no. 3 (March 2014), 109–110; Andrea Moro, "Response to Pulvermueller: The Syntax of Actions and Other Metaphors," *Trends in Cognitive Sciences* 18, no. 5 (2014): 221.
40. Moro, *Impossible Languages.*
41. Lorenzo Magrassi, Giuseppe Aromataris, Alessandro Cabrini, Valerio Annovazzi-Lodi, and Andrea Moro, "Sound Representation in Higher Language Areas during Language Generation," *Proceedings of the National Academy of Sciences* 112, no. 6 (2015): 1868–1873; Moro, *Impossible Languages*; Fiorenzo Artoni, Piergiorgio d'Orio, Eleonora Catricalà, Francesca Conca, Franco Bottoni, Veronica Pelliccia, Ivana Sartori, Giorgio Lo Russo, Stefano F. Cappa, Silvestro Micera, and Andrea Moro, "High Gamma Response Tracks Different Syntactic Structures in Homophonous Phrases," *Nature Scientific Reports* 10 (2020): 7537.
42. One such technical obstacle is the need for a device that can provide a comprehensive and less invasive recording of the electrical activity of neurons.
43. Abdus Salam, *The Unification of Fundamental Forces* (Cambridge: Cambridge University Press, 1990).

44. Chierchia, *Logic in Grammar*; Maria Vender, Diego Gabriel Krivochen, Arianna Compostella, Beth Phillips, Denis Delfitto, and Douglas Saddy, "Disentangling Sequential from Hierarchical Learning in Artificial Grammar Learning: Evidence from a Modified Simon Task," *PLOS ONE* 15, no. 5 (2020): 1–26.
45. Morris Halle and Alec Marantz, "Distributed Morphology and the Pieces of Inflection," in *The View from Building 20*, ed. Kenneth Hale and S. Jay Keyser (Cambridge, MA: MIT Press, 1993), 111–176; Richard S. Kayne, "What Is Suppletion? On **Goed* and on *Went* in Modern English," in "The Diachrony of Suppletion," ed. Nigel Vincent and Frans Plank, special issue, *Transactions of the Philological Society* 117, no. 3(2019): 434–454.
46. Andrea Calabrese, *Markedness and Economy in a Derivational Model of Phonology* (Berlin: Mouton de Gruyter, 2005); Andrew Nevins, *Locality in Vowel Harmony* (Cambridge, MA: MIT Press, 2010); Marc van Oostendorpo, ed., *Blackwell Companion of Phonology* (Oxford: Blackwell, 2011); Nicholas Allott and Deirdre Wilson, "Chomsky and Pragmatics," in *A Companion to Chomsky*, ed. Nicholas Allot, Terje Lohndal, and Georges Rey (New York: Wiley, 2021).
47. Giuseppe Longobardi and Ian Roberts, "Universals, Diversity and Change in the Science of Language: Reaction to 'The Myth of Language Universals and Cognitive Science,'" *Lingua* 120, no. 12 (2010): 2699–2703.
48. Moro, *Impossible Languages*.
49. Emil du Bois-Reymond, "The Limits of Our Knowledge of Nature," *Popular Science Monthly* 5 (1874): 17–32.

50. Moro, *I Speak, Therefore I Am.*
51. Any errors or omissions in this afterword are my responsibility, but the reduced number of them is certainly due to the constructive criticism of Noam Chomsky and the polydimensional revision of the text by Marc Lowenthal, who deleted all traces of impossible languages from my own writing. Special thanks also to Judith Feldmann for an accurate revision of my English, which made my thoughts much clearer (and the errors as well).